Scientific Proof of G-D

By Zivi Ritchie, BsEE

ISBN: 9798335972482

© copyright 2024 by Zivi Ritchie

The second book of the series: Secrets of the Torah

Buy the book from Amazon.com

Download this PDF E-book for free at
www.263672.com

Please help spread the word.

Help get this book to everyone.

It is allowed only to send the book as is, and don't omit, change, or add anything.

Help spread this message. You can buy a smaller version of this book in bulk at minimum cost to give away to friends.
Get 100 books for $2 each plus shipping. Contact us at:
Refuahoffice@gmail.com
www.Refuah.net
1-646-395-9613

Table of Contents:

Page 3. Perfect eclipse, not by chance

Page 5. Other examples of precise correlations

Page 9. Help spread the word

Page 10. The proof that humanity has been waiting for

Page 13. How we calculate probability

Page 17. Visual depiction of correlations

Page 20. Calculation of all 26 correlations

Page 27. Explanation of alphanumerical codes

Page 30. Geometric proofs of G-d

Page 31. The 7 Noahide Laws

Page 32. Secrets of science found in the Talmud

Page 33. Why this proof is irrefutable

Page 34. The future of science and education

Page 36. About the author

Scientific proof of G-D

Mathematical proof, numerical codes, script, production, and animation by Zivi Ritchie

© copyright 2024 Zivi Ritchie

Please share this with your friends

 SHARE

Visit our new website:
www.263672.com

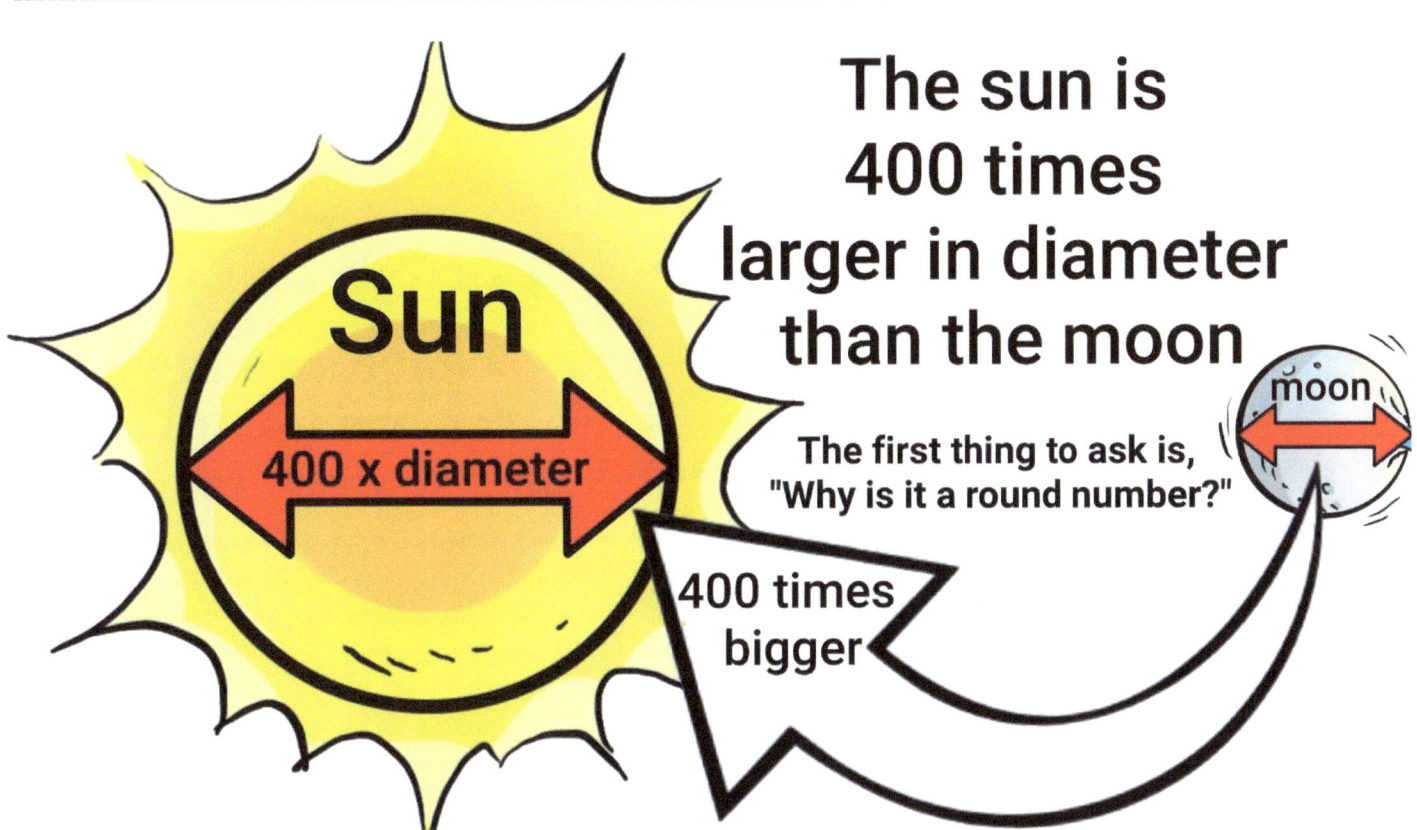

The sun is 400 times larger in diameter than the moon

The first thing to ask is, "Why is it a round number?"

400 x diameter

400 times bigger

The distance between the sun and the earth is also 400 times more than the distance between the earth and the moon

moon

400 times farther away

400 x distance than the moon

Sun

Earth

The next thing to ask is, "Why is this also the same round number, 400?"

This is why the moon covers the sun perfectly during a solar eclipse

Sun

moon

Solar eclipse

Because the ratio between the sun's distance to earth and the moon's distance to earth is the same as the ratio between the sun's diameter and the moon's diameter.

Wow, that coincides perfectly! This looks like it was created this way on purpose!

1. Are there other special ratios and numbers like these?

2. What's the probability of the numbers being a round 400?

3. What is the probability of the ratios being the same?

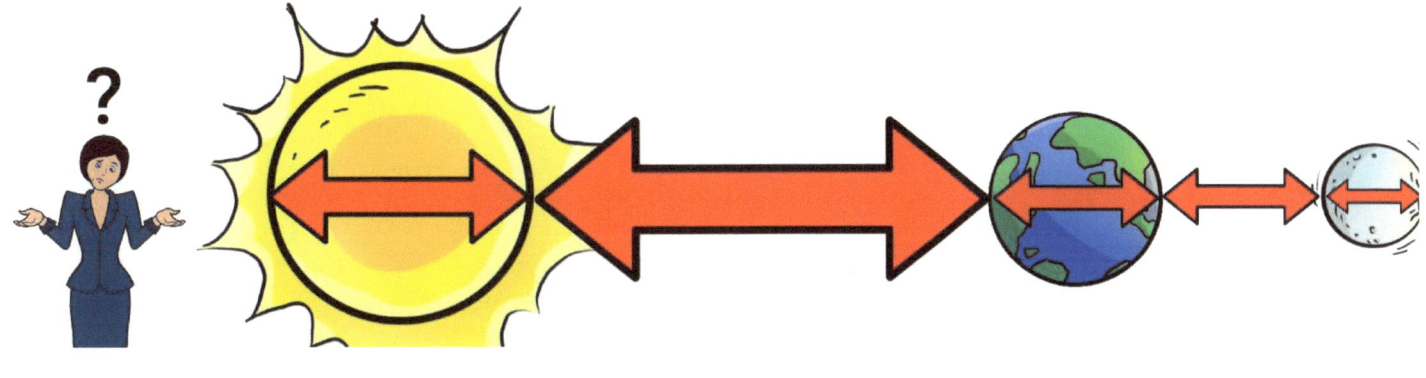

Yes, there are other ratios and numbers that are special:

The Moon is 30 Earth Diameters Distant from the Earth...

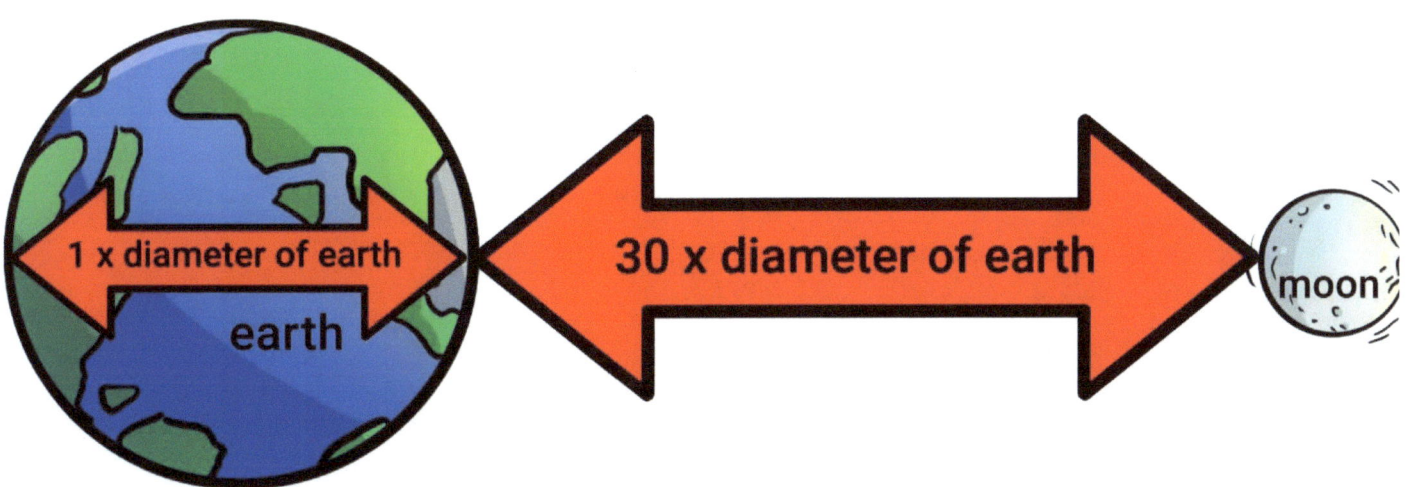

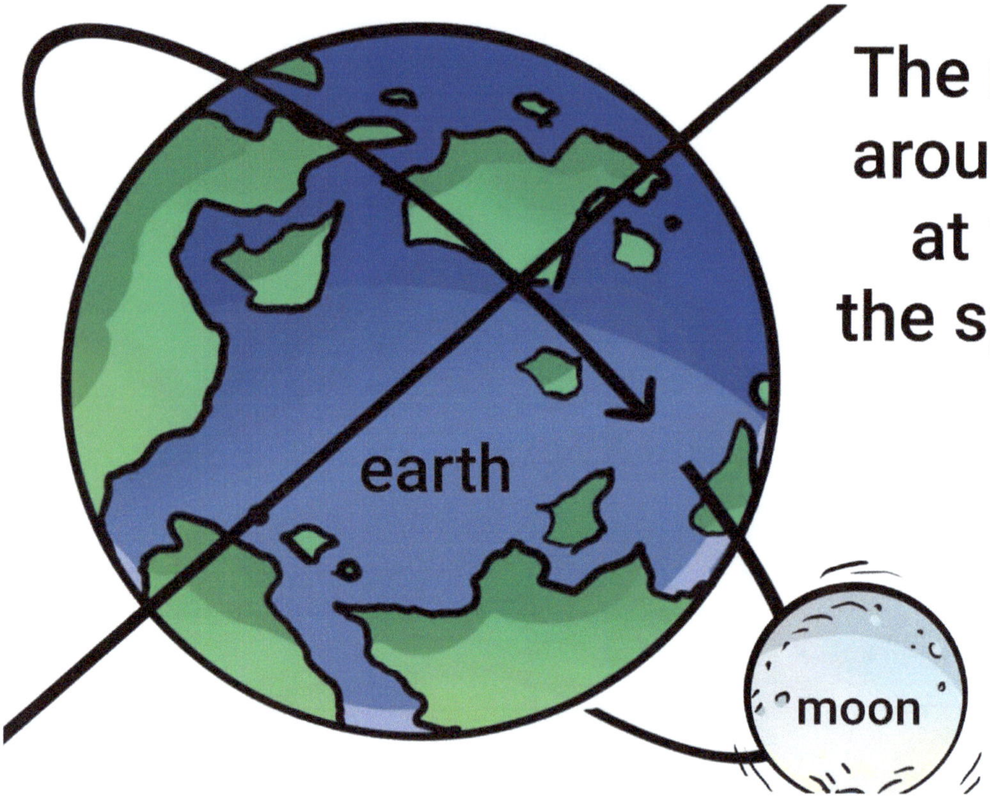

The moon orbits around the earth at 1/300,000 the speed of light.

The moon orbits the earth in a period of about 30 times the amount of time of earth's rotation.

The sun rotates at different time periods near its poles and near its equator, but the average time of the sun's rotation is the same time as the time of the moon's orbit around the earth.

The earth orbits around the sun at 1/10,000 the speed of light.

The rotational velocity of the earth is 100 times the rotational velocity of the moon.

What are the odds that all of the 26 possible ways
to correlate the precise measurements of
the sun, earth, and moon,
using metrics of time, distance, and the speed of light,
would coincide to each other so perfectly by random chance?
The probability of this being random is:
1 / 1,080,000,000,000,000,000,000

This is proof that the world was created by G-d.

Order the book
By Zivi Ritchie
from Amazon.com

Eclipse!
Mathematical Proof that the
World was Created by G-d.

or

Download the e-book for free from
www.263672.com

Once the proof in this book is spread to the world,
there will not be anyone who will
claim to be an atheist or agnostic
and still be able to claim that they are scientific.

Be a part of this great change in the world
by sharing this book and video with your friends.

This book has mathematical proof
that the world was created by G-d.
This is the first time in the history of the world
that this proof was documented
using classical probability mathematics.

Through the most recent exact
measurements of our solar system,
we can now see that the world
was created by intelligent design.

The chances of all of the precise metrics
of the sun, moon, and earth
to just happen to coincide with these whole digits
without being designed on purpose this way, is
1 / 1,080,000,000,000,000,000,000
which is conclusively impossible.
This is one chance in a thousand billion billion!

To put this into perspective:
The lifetime risk of dying from a car accident is about 1/100.
People don't refrain from using cars because they
look at a 1/100 chance to be too small to consider.
People obviously don't think that a
1 / 1,080,000,000,000,000,000,000
chance is anything to consider.

From now on, it is no longer just a matter of opinion or belief
that the world was created,
but it is now a proven scientific fact
that the world is not just a random occurrence.

For the past approximately two hundred years,
atheists have said that if someone would
just show them one scientific proof that G-d exists,
they will believe in G-d.
Well, finally, here it is in this book,
the scientific proof that they asked for.

Now let's calculate
what is the exact probability
of this being random.

Let's take this first example:

The sun is exactly 400.800 times larger in diameter than the moon

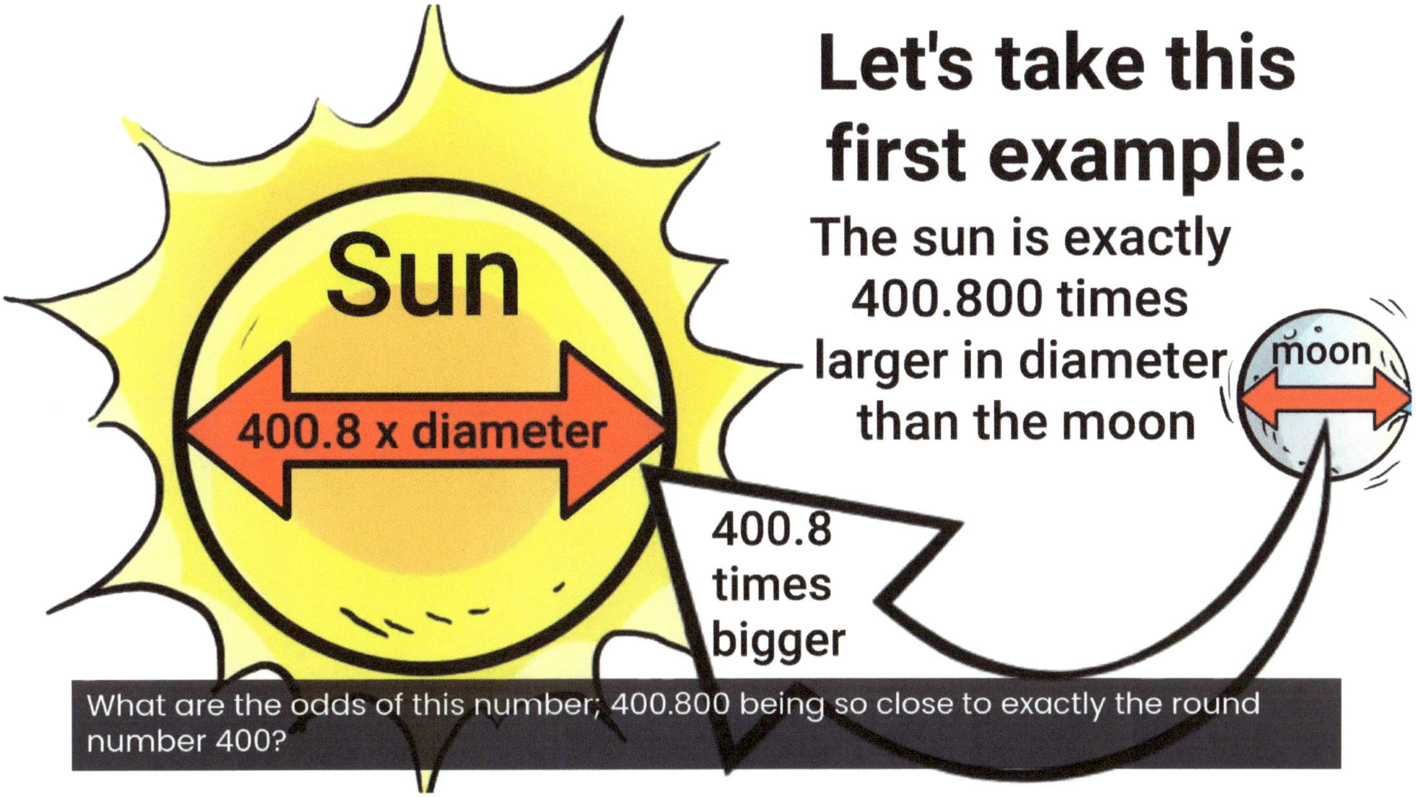

Sun — 400.8 x diameter

400.8 times bigger

What are the odds of this number; 400.800 being so close to exactly the round number 400?

This is how we calculate the probability. We ask: What are the odds of a random number being close to a whole number, such as 1, 2, 3... or 10, 20, 30... etc...

As opposed to the odds of the random number being as far away as possible from the round number, for example 1.5, 2.5, 3.5 ... or 15, 25, 35 ... etc...

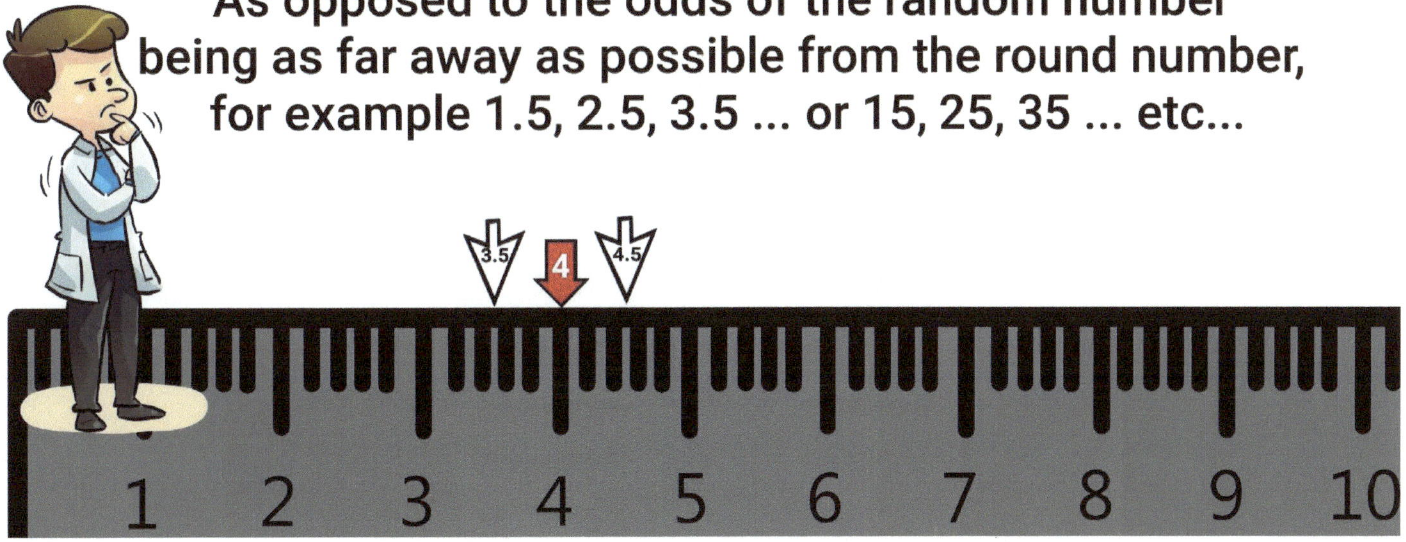

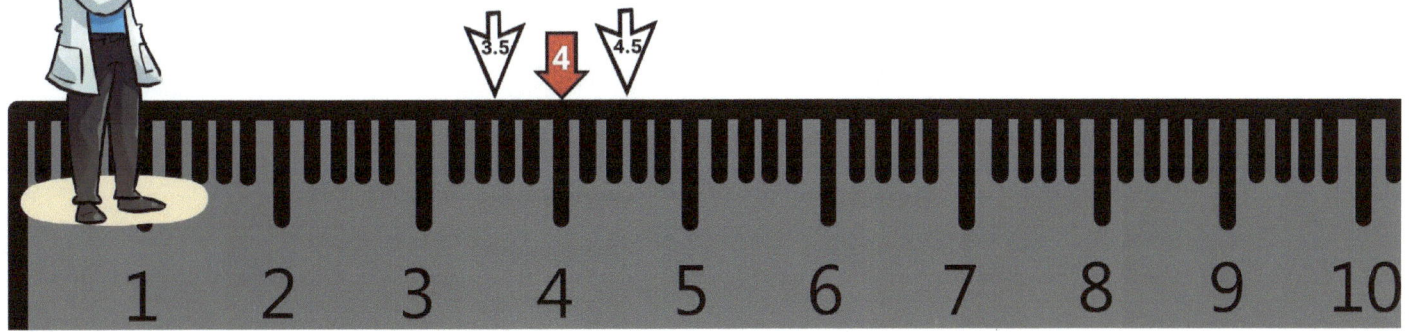

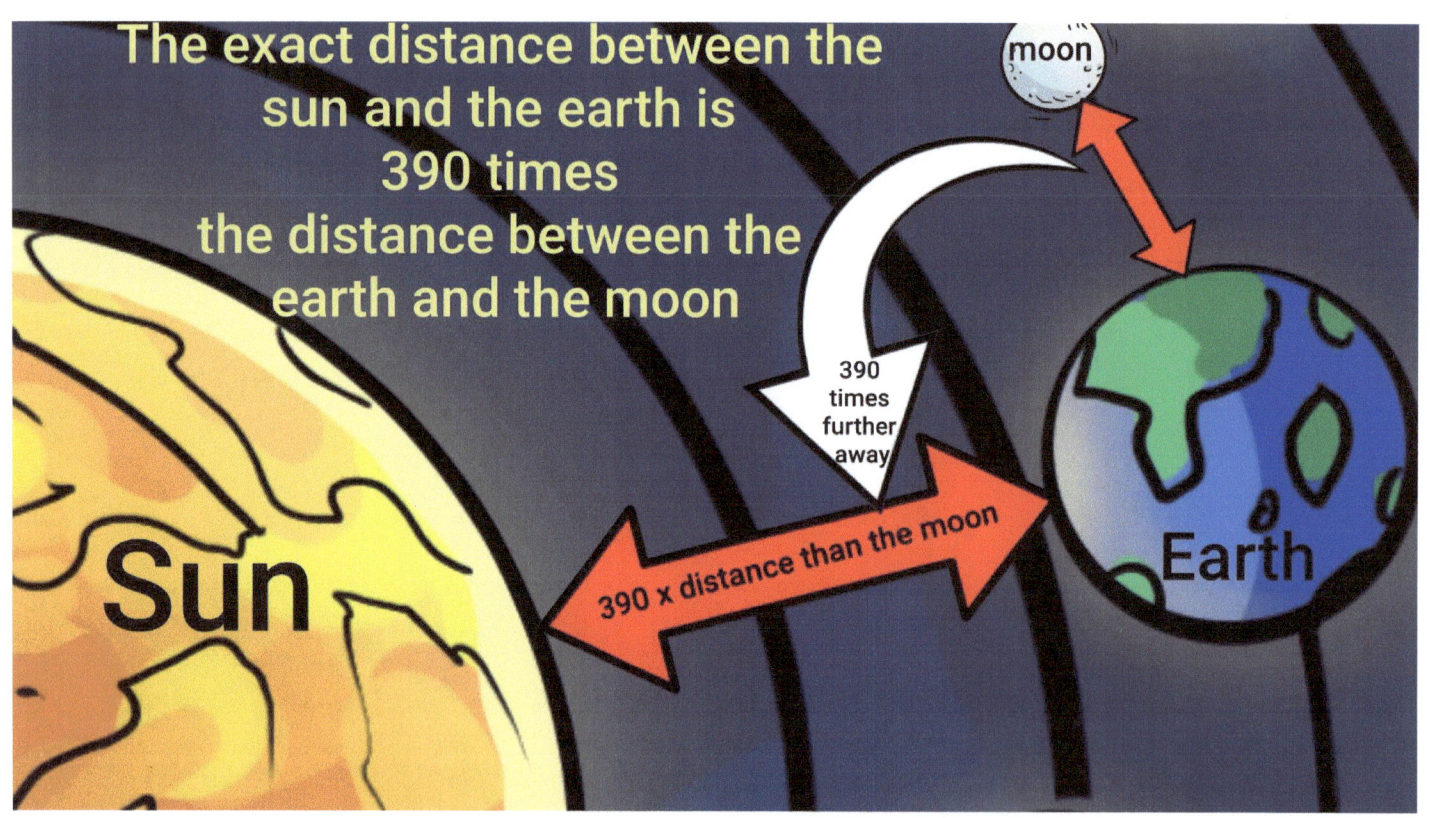

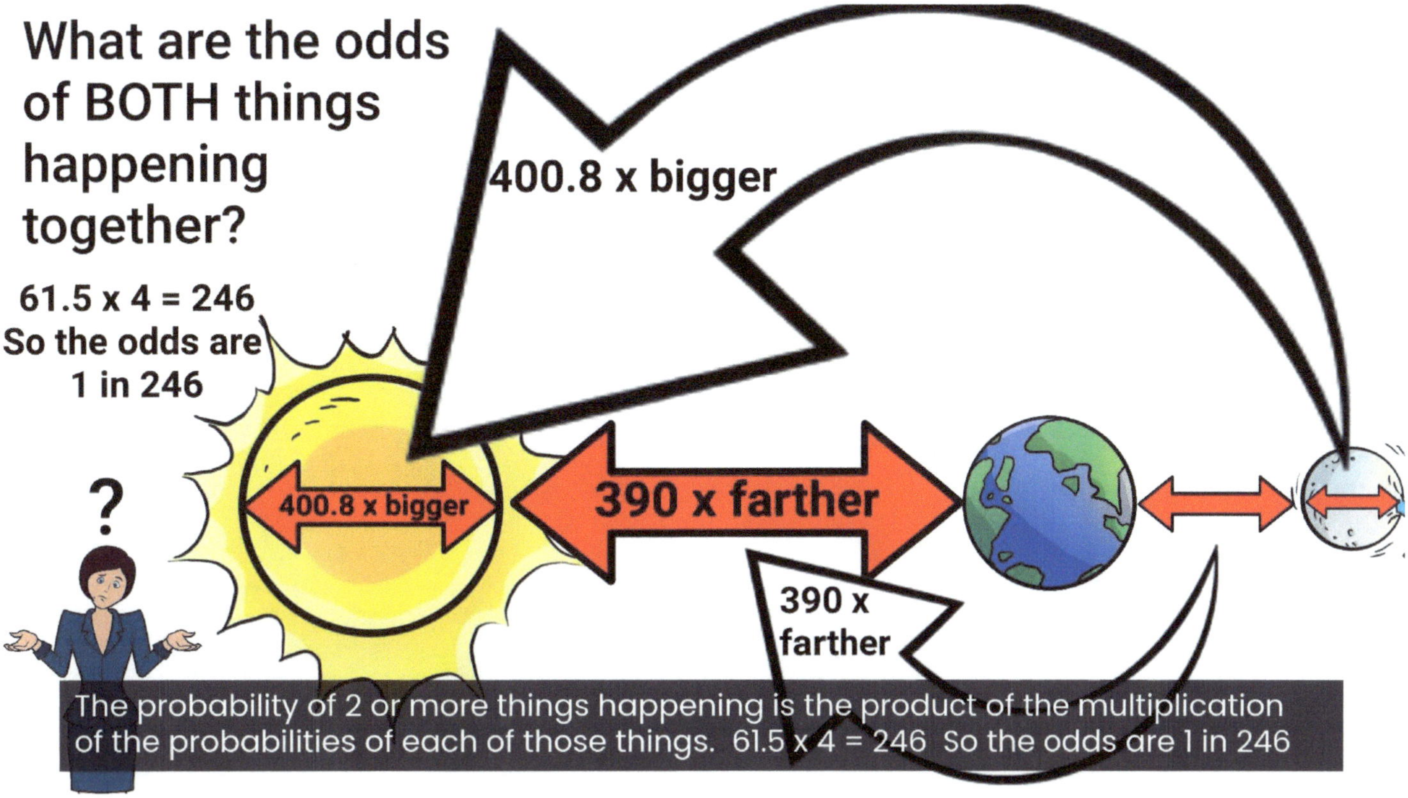

What are the odds of ALL of these 8 things happening together?

1. The size of the moon compared to the size of the sun. The odds are 1 in 61.5

2. The size of the earth compared to the size of the sun. The odds are 1 in 4.3

3. The size of the moon compared to the size of the earth. The odds are 1 in 0.045

4. The distance the orbit of earth compared to the distance of the orbit of the moon. The odds are 1 in 4

In order to properly calculate the probability of being random, we need to check all of the possibilities.

So we check everything:
A. All of the body diameters compared to body diameters.
B. All of the orbit radius compared to the orbit radius.
C. All of the orbit radius compared to the diameter of the orbiting body, and to the diameter of the body being orbited.

5. The distance of the orbit the moon around the earth compared to the diameter of the earth. The odds are 1 in 26

6. The distance of the orbit the moon around the earth compared to the diameter of the moon. The odds are 1 in 3.7

7. The distance of the orbit the earth around the sun compared to the diameter of the sun. The odds are 1 in 5.7

8. The distance of the orbit the earth around the sun compared to the diameter of the earth. The odds are 1 in 1.9

The odds of all these 8 things happening is:
61.5 x 4.3 x 0.045 x 4 x 26 x 3.7 x 5.7 x 1.9 = 45,000

The odds are 1 in 45,000

This means that there is only a 1 in 45,000 chance that this happened randomly.

This is the first part of the proof that the world was created by G-d!

These are all of the possible ways to make correlations between the diameters of the spheres orbiting and being orbited, and the distance between them.

**Did you notice
that all of the numbers of the correlations
between the diameters of the bodies
and the orbits
are multiples of 1, 10, 100...
or 3, 30, 300...
or 4, 40, 400...
What is the probability
of that being randomly by chance?**

**These numbers are repeatedly mentioned in the Torah, Hebrew Bible:
There are 10 commandments.
The tabernacle had measurements of 10 cubits and 30 cubits.
Esau came with 400 men to fight Jacob.
There were 40 days and 40 nights of rain during the Flood of Noah. And more...**

**Imagine that you saw a fence with 10 slats.
And you saw arrows that were shot into it.
And the arrows were grouped together
in 3 clusters like this: Slat 1, slat 3, and slat 4.
Would you think that this was a random spread from a volley?
Or would it be obvious that these arrows were aimed on purpose!**

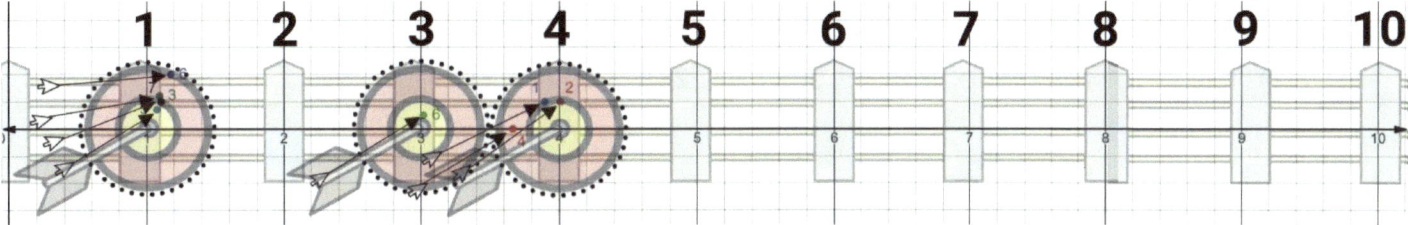

**This graph shows the 8 ratios.
They all are tightly grouped around
multiples of numbers 1, 3, or 4... 10, 30, 40... etc...**

And imagine that you have 10 bins numbered 1 - 10, and you place the 10 bins next to each other in a box. And you randomly throw 8 balls into the box so that they randomly fall into any of the 10 bins.

(In a way that they can't fall outside of the bins.)

This shows you in a visual picture how the ratios of the diameters and distances of the sun, moon and earth are not random.

And we see that all of the 8 balls fell into bins numbered 1, or 3, or 4.
(And they didn't fall into any of the other bins.)
What is the probability of this happening by chance?

First, we need to take into consideration a few mathematical things:

1. When we calculate the ratios, we always divide the larger number by the smaller number, and then we reduce to one decimal place.
This makes it always a number above 1 and below 10, but never between 0 to 1.
And so there are actually only 9 possibilities instead of 10.

2. We will only count the probabilities that are repeatedly recurring in the same numbers 1, or 3, or 4.

So the probability is: 1/3 x 1/3 x 1/3 x 1/3 x 1/3 = 1/243

The probability of the ratios repeatedly being in the range of 1, or 3, or 4, is
1 / 243

What is the probability of both happening?

A. The odds of the 8 numbers being so close to a round number; 1, 2, 3... or 10, 20, 30... etc...
= 1 in 45,000

AND

B. The probability of all of the 8 numbers falling repeatedly into the numbers 1, or 3, or 4.
= 1/243

The chances of both happening, that the 8 numbers are both close to a round number, and that they are all repeatedly in the range of 1, or 3, or 4, is:

1/45,000 x 1/243 = 1/10,944,858

The total probability is 1 / 10,944,858

This is the same probability of picking the winning numbers for the million dollar prize of the powerball lottery.

The odds of these first correlations of the 8 distances being randomly by chance is less than one in ten million!

1/10,000,000

This is Proof that G-d created the world!

There is a limit to what people will accept as being random.
"Beyond reasonable doubt is considered to be
one of the highest standards of proof in criminal law.
It is placed in the certainty range of 98 or 99 percent."
So when something has a 1/10,000,000 chance
it would obviously always be considered beyond any doubt.

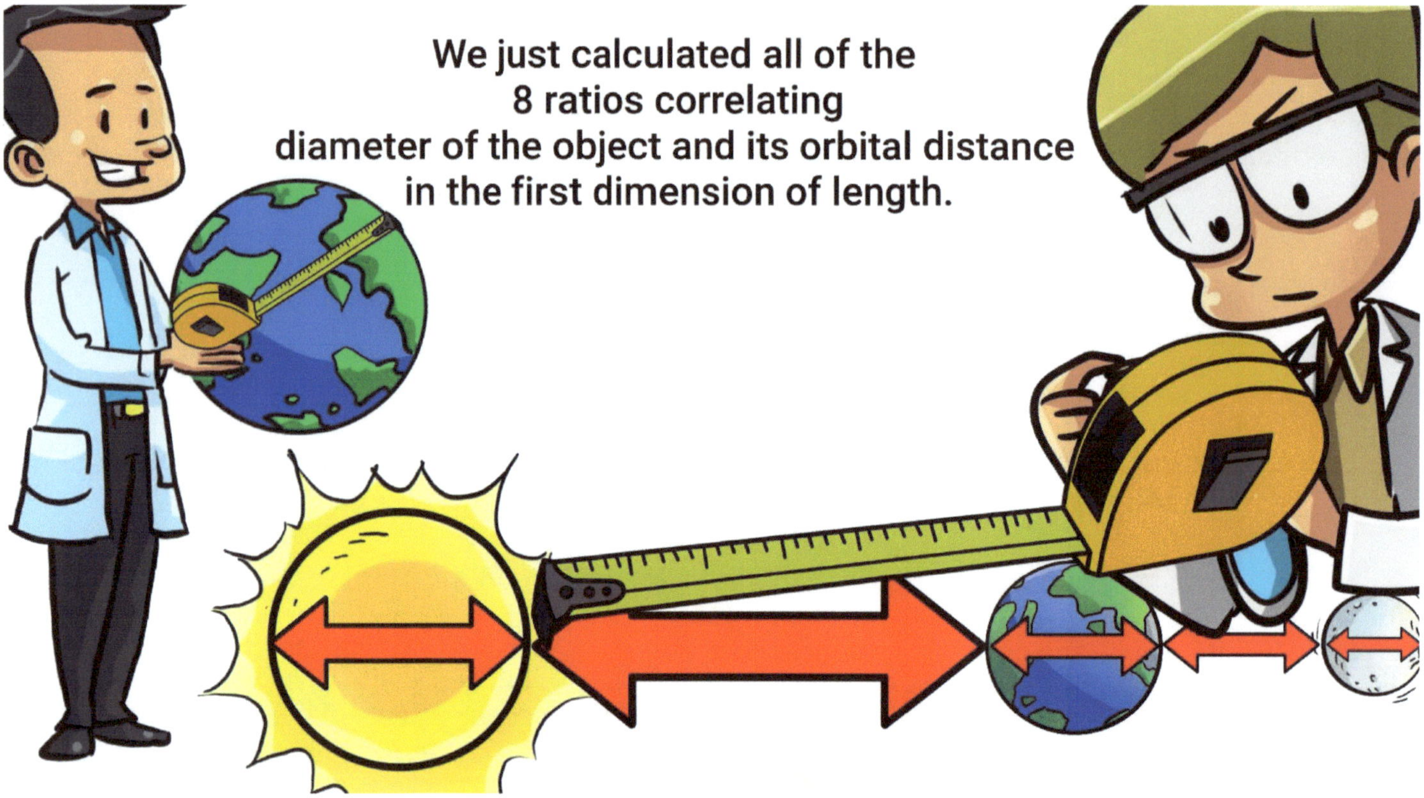

We just calculated all of the 8 ratios correlating diameter of the object and its orbital distance in the first dimension of length.

Now let's measure all of the ratios that are in the dimension of time!

1. Time period of rotation.
2. Average rotational velocity
3. Speed of orbital velocity

9. The moon orbits the earth in a period of 29.53 times the time of earth's rotation. The odds are 1 to 9.6

10. The sun has different rotational time periods near its poles and near its equator, but the average rotation time period of the sun is (25.05+34.4)/2 = 29.725 days. The odds are 1 to 17

11. The average time of the sun's rotation is 1.0066 times the moon's orbit around the earth. The odds are 1 to 75

12. The orbital velocity of the earth is 29.14 times the orbital velocity of the moon circling the earth. The odds are 1 to 4.8

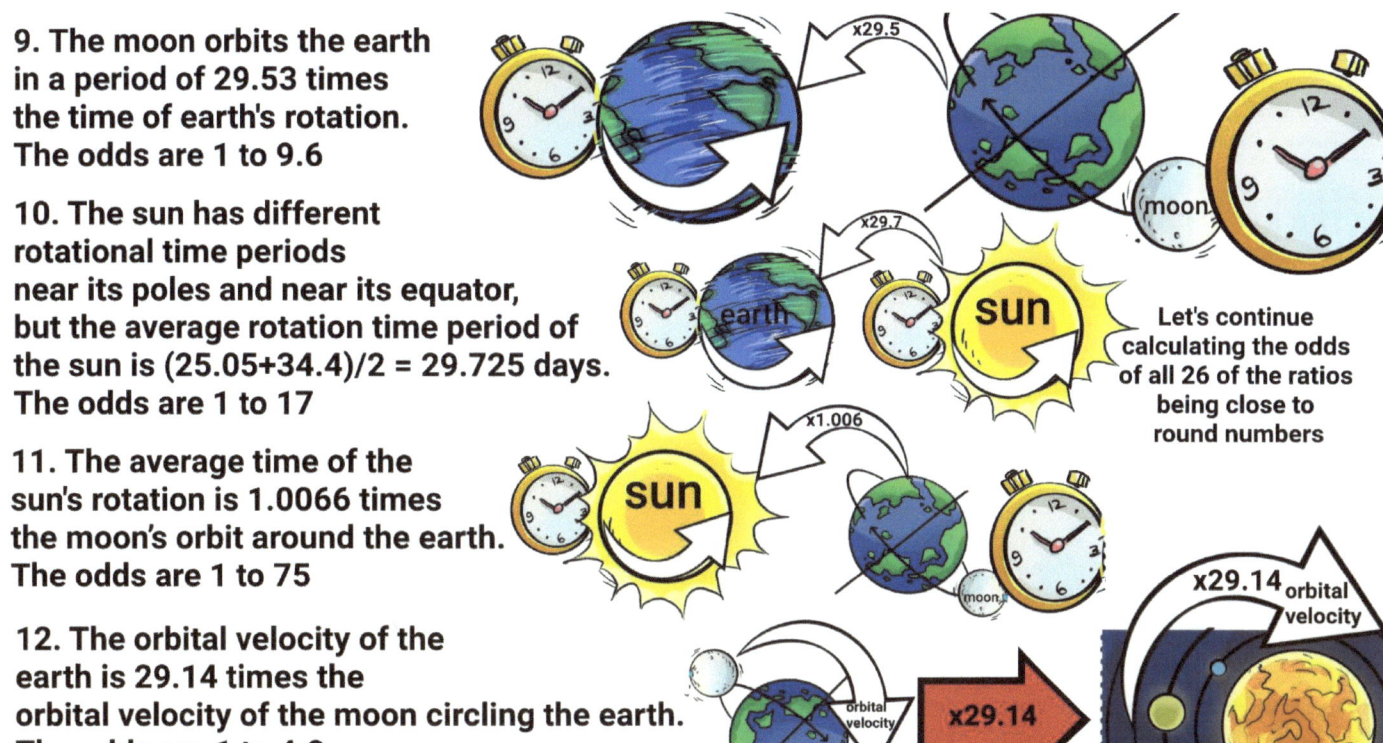

Let's continue calculating the odds of all 26 of the ratios being close to round numbers

---

13. The rotational velocity of the earth is 100.5 times the rotational velocity of the moon. The odds are 1 to 95

14. The sun rotates with different time periods near its poles and near its equator, but the average rotational velocity of the sun is 3.7 times the average rotational velocity of the earth. The odds are 1 to 0.73

15. The average rotational velocity of the sun is 373 times the moon's average orbital velocity. The odds are 1 to 0.85

* I consistently calculate all the average values by simply using the formula: (largest value + smallest value)/2.
I calculate average rotational velocity of a spherical body, using this formula: (Velocity pole + Velocity equator)/2.
And since the suns rotational time period is fluid and takes more time at the poles,
so I also factor in the sun's average rotational time period using this formula:
((Velocity equator + Velocity pole)/2) x (((Time period near equator + Time period near poles)/2)/Time period near poles)

16. The orbital velocity of the earth is 128 times the average rotational velocity of the earth.
The odds are 1 to 0.78

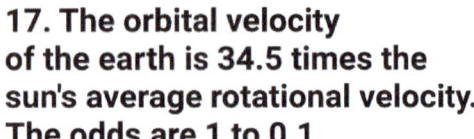

17. The orbital velocity of the earth is 34.5 times the sun's average rotational velocity.
The odds are 1 to 0.1

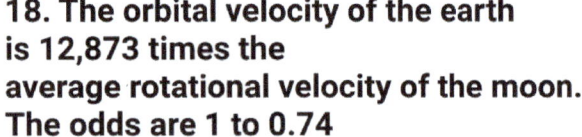

18. The orbital velocity of the earth is 12,873 times the average rotational velocity of the moon.
The odds are 1 to 0.74

---

19. The moon's orbital velocity is 4.4 times the earth's average rotational velocity.
The odds are 1 to 0.26

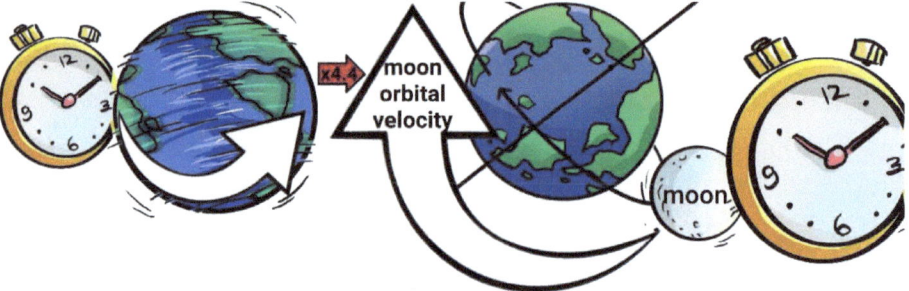

20. The moons orbital velocity is 1.2 times the sun's average rotational velocity.
The odds are 1 to 1.7

21. The moons orbital velocity is 442 times the earth's average rotational velocity.
The odds are 1 to 0.2

Now let's compare all of the velocities to the speed of light!

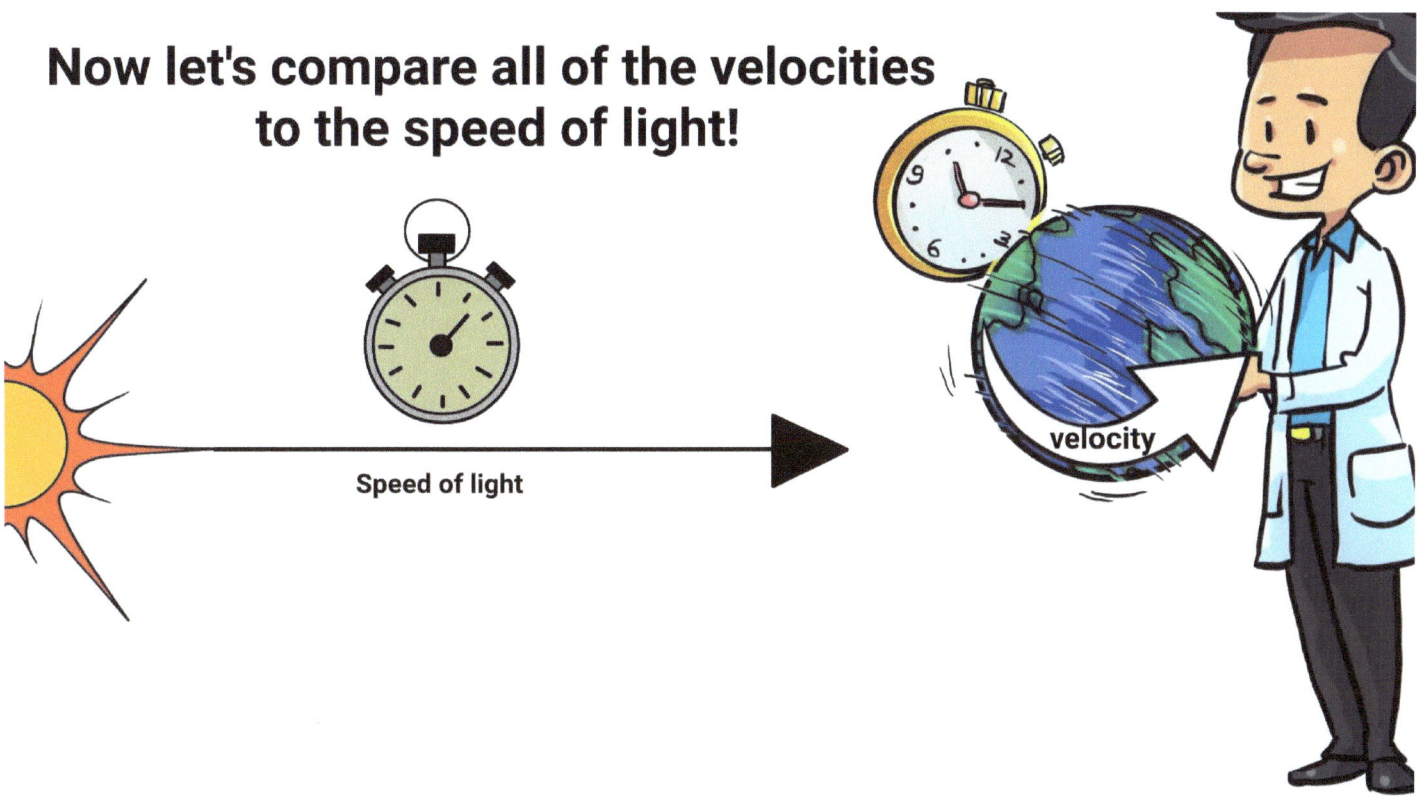

22. The speed of light is 129,583,946 times the average rotational velocity of the moon. The odds are 1 to 0.7

23. The speed of light is 1,289,153 times the average rotational velocity of earths. The odds are 1 to 0.73

24. The speed of light is 347,464 times the average rotational velocity of the sun. The odds are 1 to 0.053

25. The speed of light is 293,339 times the orbital velocity of the moon. The odds are 1 to 6.5

26. The speed of light is 10,066 times the orbital velocity of the earth. The odds are 1 to 75

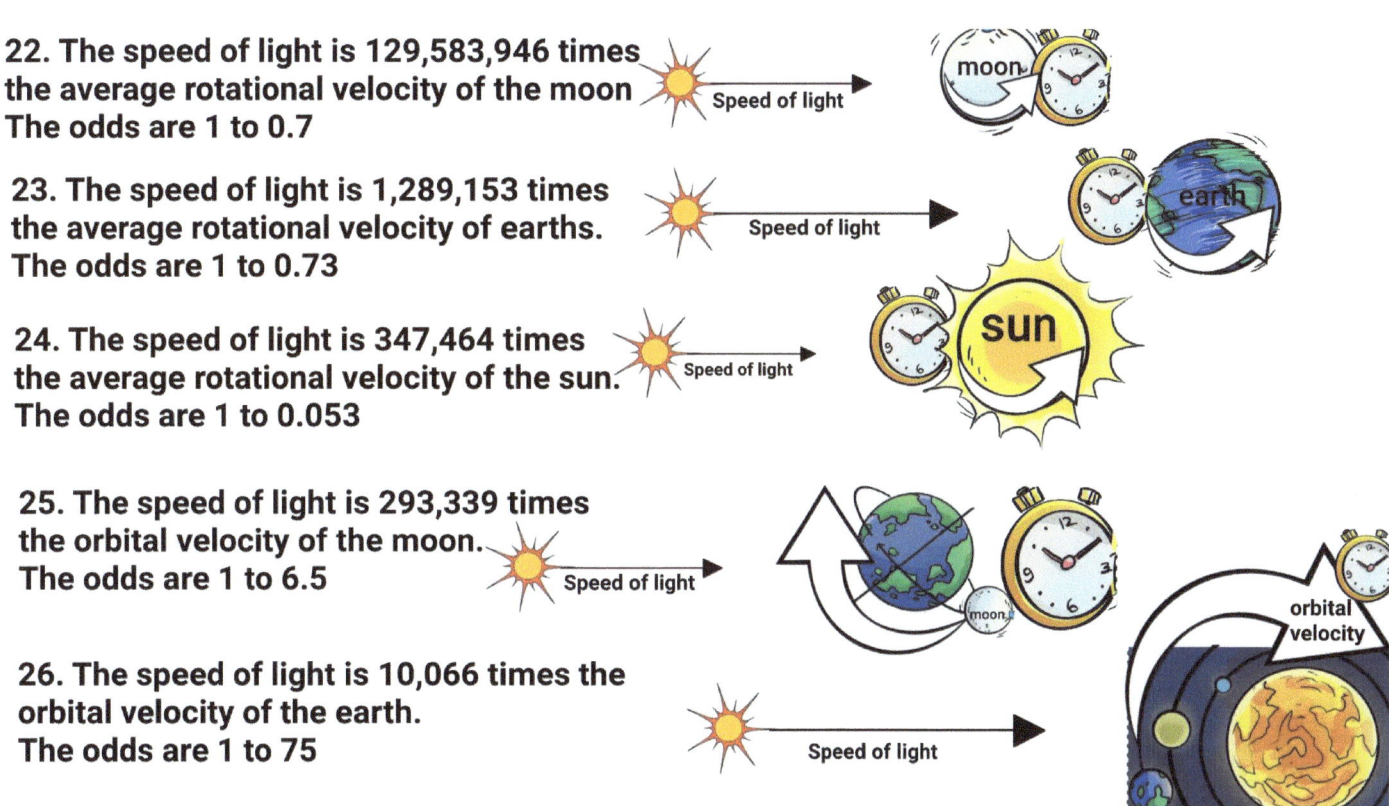

The odds of all of the 26 ratios being close to whole numbers is:

1 / (61.5 x 4.3 x 0.045 x 4 x 26 x 3.7 x 5.7 x 1.9 x 9.6 x 17 x 75 x 4.8 x 95 x 0.73 x 0.85 x 0.78 x 0.1 x 0.74 x 0.26 x 1.7 x 0. 2 x 0.7 x 0.73 x 0.053 x 6.5 x 75)
= 1 / 11,500,000,000

1 / 11,500,000,000

That is 1000 times less likely than picking the winning numbers for the one million dollar powerball lottery prize!

But even more surprising, ALL of the 26 possible correlations are in the range of 1, or 3, or 4

What is the probability of that happening?
1/3 x 1/3
= 1/(3^23)

The probability is 1/94,143,178,827

The chances that all 26 correlations are repeatedly in the range of 1, or 3, or 4 is about one in one hundred billion!

The chances of the ratios being so close to whole numbers,
AND all 26 of them being in the range of numbers 1, or 3, or 4 is:

1/11,500,000,000 x 1/94,143,178,827 = 1/1,080,000,000,000,000,000,000

The total odds of this being random is
one to a thousand billion billion!

1 / 1,080,000,000,000,000,000,000

That is as unlikely as someone picking the
winning numbers of the million dollar
powerball lottery prize 3 times in a row!

Now lets look at this graph of the fence again,
but this time, with all 26 points.
We see that all 26 of them are
grouped inside the range of 1, or 3, or 4,
and none of them are in the ranges of 2, 5, 6, 7, 8, or 9.
Would you think that this could possibly be random?
If it was random, then the points would be spread out between 1 to 10.
YOU CAN SEE WITH YOUR OWN EYES THAT THE WORLD IS NOT RANDOM!!!!!

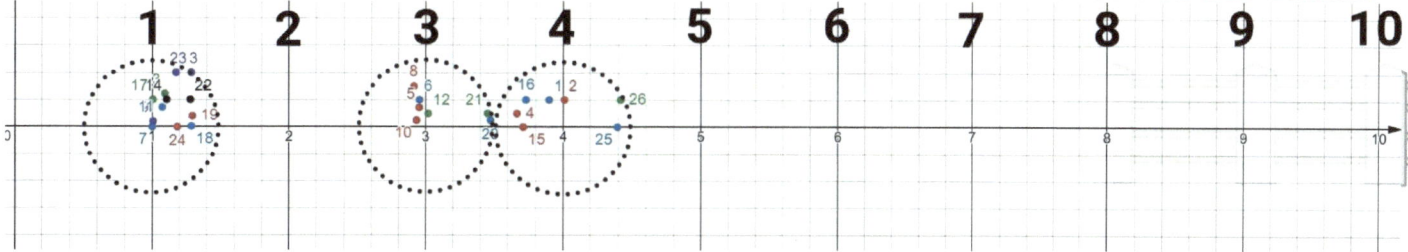

This is for the people who say that they want to see
visual proof of G-d in order to believe it.
You are looking at it right now!

Let's look again at the diagram of the 10 bins in the box, but now with with all 26 balls. Imagine that you have 10 bins numbered 1 - 10, and you place the 10 bins next to each other in a box. And you randomly throw 26 balls into the box so that they randomly fall into any of the 10 bins. If those scientists who claim that the world is random and not created were correct, then the balls (representing the ratios) should fall more or less evenly into all of the 10 bins, with 2 or 3 balls in each bin. But what do we see instead? ALL of the 26 balls are in just 3 bins, and NONE of the balls are in the other 7 bins!

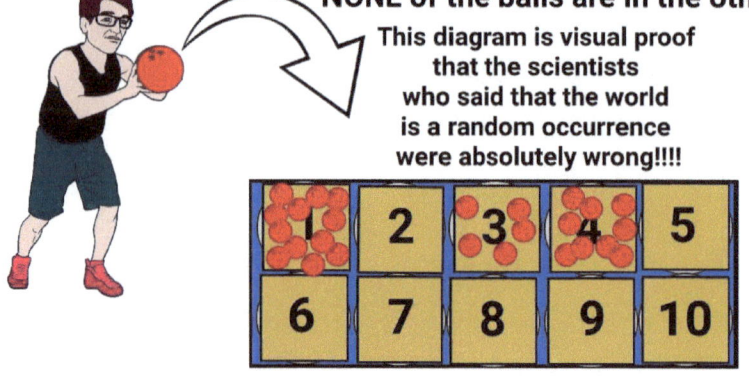

This diagram is visual proof that the scientists who said that the world is a random occurrence were absolutely wrong!!!

Every single one of the 26 ratios of the measurements between the sun, earth and moon are all in the range of either 1, or 3, or 4. And NONE of them are in the range of any of the other numbers!

DOES THIS IN ANY WAY LOOK TO YOU LIKE THE 26 BALLS ARE RANDOMLY SPREAD OUT BETWEEN ALL 10 BINS?? ABSOLUTELY NOT !!!

If someone said that he will win
the one million dollar powerball lottery prize
3 times in a row by only purchasing three lottery tickets,
no one would believe that it would happen.
And if he then did actually win all 3 times,
nobody would believe that it was by chance.
They would investigate to see how he rigged the system.

So too, no one can believe that the world was formed by chance,
now that we see these absolutely unlikely
precisely correlating measurements
of the sun, earth, and moon.

This is proof that G-d created the world!

What is the meaning and importance of the numbers 3, 1, and 4?

This is an alphanumeric code that spells G-O-D in Hebrew letters.

G - 3 - ג The 3rd letter of the Hebrew alphabet is Gimmel, that has the sound G.

O - 1 - א The 1st letter of the Hebrew alphabet is Alef, that also has the sound O.

D - 4 - ד The 4th letter of the Hebrew alphabet is Daled, that has the sound D.

This is the modern Hebrew spelling of the English word for G-D.

This is also the way it is in Yiddish,
because the Yiddish language uses Hebrew alphabet,
and integrates words from numerous languages,
including mostly German, Hebrew, and more recently, also some English.

The proof in this book reveals how science and math prove that
the world displays the existence of G-D
in terms that the entire world can see and understand,
by using a language that combines Hebrew letters
with the universally used language of science which is English,
through the use of gematria (numerical values of letters).
This is why we are seeing the gematria numerical value of
the word for G-D in English, spelled with Hebrew letters.

The name for G-d in Hebrew
that is written on the Mezuza parchment of the doorposts is:

ש – ד – י

The 'reduced to one decimal point' gematria numerical value is:

ש = 3

ד = 4

י = 1

This is how the gematria of 'reduced to one decimal point' works:
ש = 300 => 3   ד = 4 => 4   י = 10 => 1

The Name of G-d in Hebrew is:
Y - H - V - H

Y = 10 = י
H = 5 = ה
V = 6 = ו
H = 5 = ה

The numerical value is:
10 + 5 + 6 + 5 =
26

=

ג = 3 , 3X3 = 9
א = 1 , 1X1 = 1
ד = 4 , 4X4 = 16

The numerical value is:
9 + 1 + 16 =
26

This next gematria was told to me by my friend,
Reb Yankala Shemesh,
who discovered it himself:

G - is the 7th letter of the English alphabet
O - is the 15th letter of the English alphabet
D - is the 4th letter of the English alphabet

The numerical value of name of G-d in English;
G = 7, O = 15, D = 4
7 + 15 + 4 = 26
has the same numerical value of the name of G-d in Hebrew, Y-H-V-H:
Yud = 10, Hei = 5, Vav = 6, Hei = 5
10 + 5 + 6 + 5 = 26

The gematria (G=7 + o=15 + d=4 = 26 = Y-H-V-H) was taught to me by Reb Yankala Shemesh. All of the other correlations, mathematical proofs, geometric proofs, gematrias, and teachings in this video, book, and website are my own discovery, unless specifically stated otherwise.
(Zivi Ritchie)

Rabbi Elazar said that Rabbi Chanina said:
One who says something in the name of the one who said it, brings redemption to the world. As it says in the Talmud, tractate Esther B' 22, "And Esther told the king in the name of Mordechai."
ואמר רבי אלעזר אמר רבי חנינא: כל האומר דבר בשם אומרו מביא גאולה לעולם, שנאמר -אסתר ב', כב- "ותאמר אסתר למלך בשם מרדכי".

You can help spread this message to the world by sending this video link to your friends.
This material is copyright © 2024 by Zivi Ritchie.
That being said, I give permission to everyone to forward and share this material in its entirety
by file, or video, as long as you
do not change anything, or add anything, or omit anything of this material.
You must include all of the links back to my websites.
By giving credit where credit is due,
you are bringing redemption to the world, and that is our goal.
Zivi Ritchie - Phone: 1-646-395-9613 coach@coach613.com www.263672.com Refuah.net

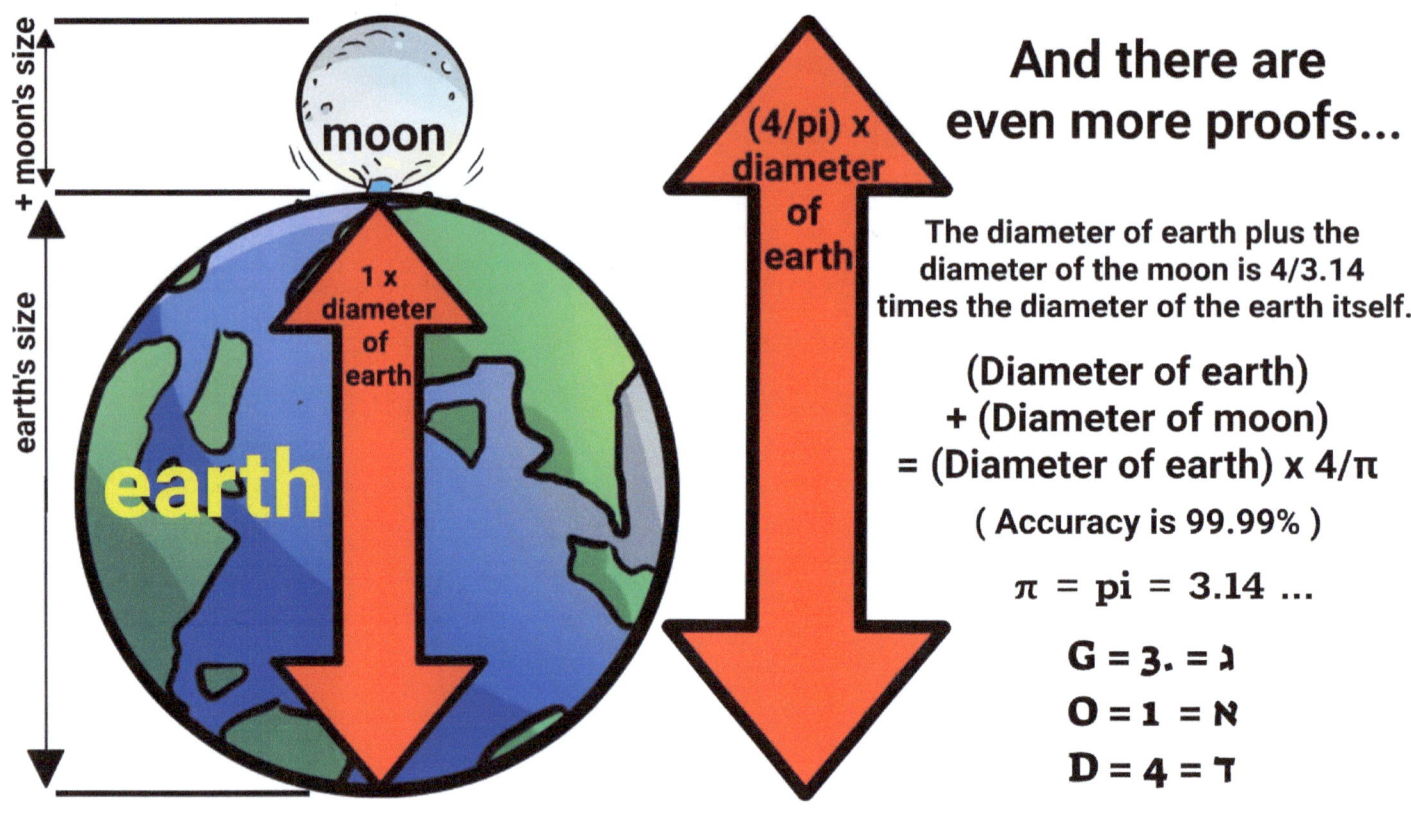

And there are even more proofs...

The diameter of earth plus the diameter of the moon is 4/3.14 times the diameter of the earth itself.

(Diameter of earth)
+ (Diameter of moon)
= (Diameter of earth) x 4/π

(Accuracy is 99.99%)

π = pi = 3.14 ...

G = 3. = ג
O = 1 = א
D = 4 = ד

The distance from the earth to the moon is 4/pi lightseconds.

(Accuracy is 99.3%)

314 = 10 + 4 + 300 = ש - ד - י

pi = π = 3.14 ...

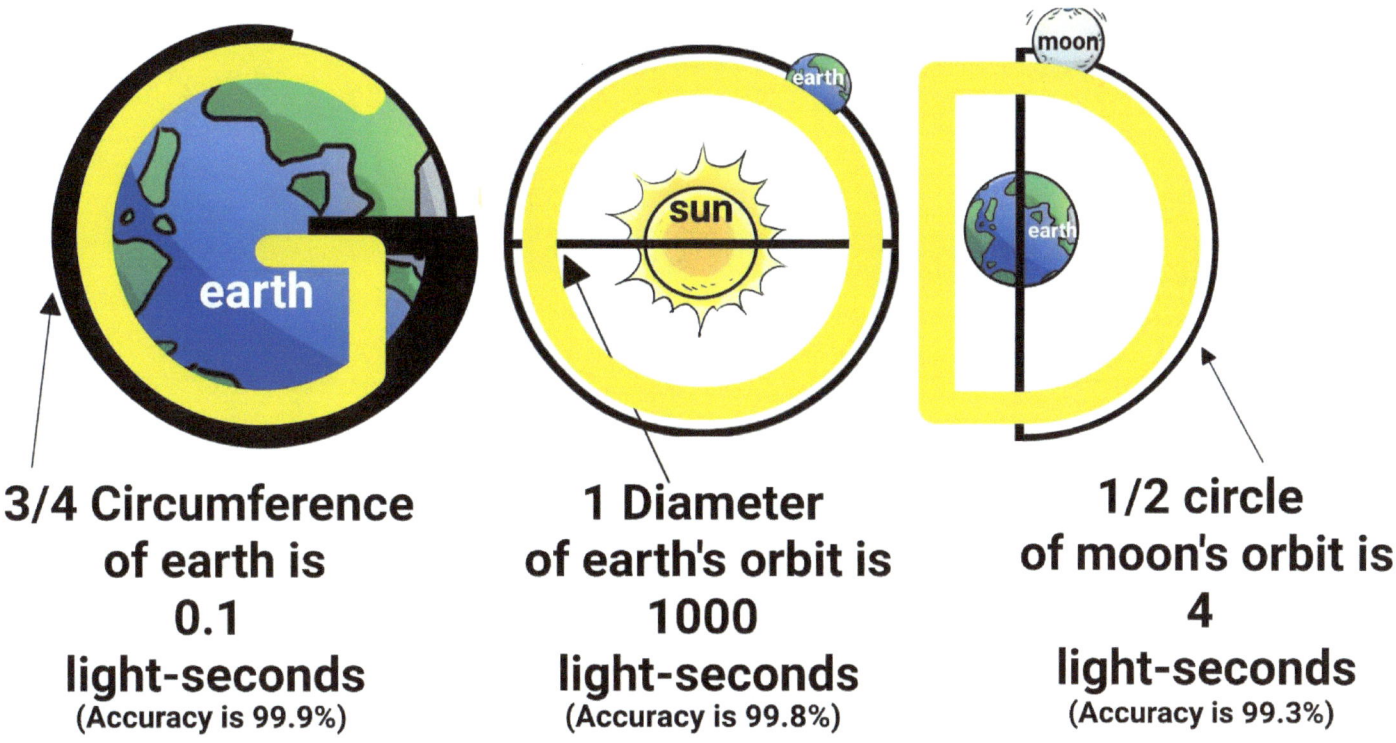

Help spread this message!   Phone: 1-646-395-9613   coach@coach613.com   www.263672.com

---

So now that we know this, what do we do now?
Everyone in the world is required to keep the 7 Noahide laws.
These laws were communicated by G-d
to Adam and Noah, ancestors of all human beings.
The 7 Noahide Laws are rules that all of us must keep,
regardless of who we are or from where we come.

1. Do not profane G-d's Oneness in any way.
2. Do not curse your Creator.
3. Do not murder.
4. Do not eat a limb of a still-living animal.
5. Do not steal.
6. Adultery with a married woman, incest,
and male homosexual relations are forbidden.
7. Establish courts of law and ensure justice in our world.

The Three asteroid belts are:
1. The Main Asteroid belt.
2. The asteroid belt in the orbit with Jupiter
3. The Kuiper belt beyond Neptune
In the Talmud, [Tractate Brachot 58B-59A]
They are call Kima, Aysh, and Ksil.
The asteroid belt of Jupiter has 2 main parts:
1. The 'Greeks' that orbit ahead of Jupiter, is called 'Reisha degla' The head of the calf.
2. The 'Trojans' that follow behind Jupiter in its orbit, is 'Zanav taleh' The tail of the lamb.
The Talmud explains how G-d brought asteroids from the asteroid belts to make the Flood of Noah.
From some time after the Talmud was written until I just recently deciphered this,
it was not interpreted correctly because we did not know that the asteroid belts existed, so they translated it incorrectly as meaning stars instead of meaning asteroids.
And they described it as 3 constellations of stars instead of as the 3 asteroid belts.

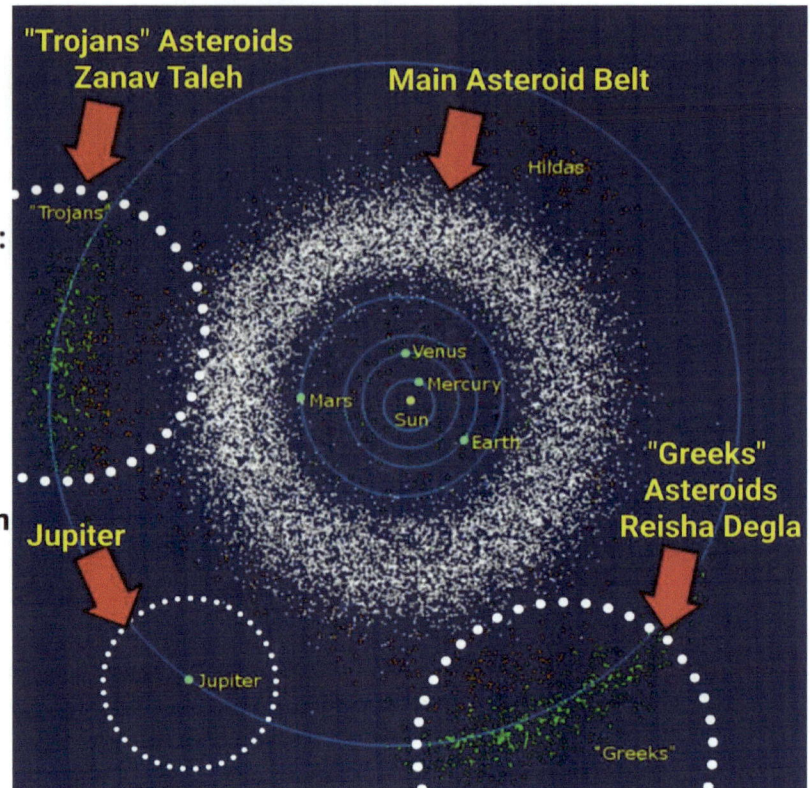

The asteroid belts were only discovered relatively recently because the asteroids are too small for people to see without special modern equipment.
How did the Rabbis writing the Talmud about 2000 years ago know about it?
[Talmud Tractate Brachot 58B-59A]
This is one of the places in the Talmud where it says "Gamiri", meaning that they know this scientific information because G-d taught it to Moses on Mount Sinai together with all of the Oral Law,
and Moses taught it to the Jewish People when he taught them all of the Oral Law.
This is proof that the Talmud and the Oral Law of the Jewish People was received from Moses who received it from G-d on Mount Sinai.

The speed off light is 299,792,458 [m/s]

A 'chelek' is 1/18 of a minute = 3.333 [seconds]
It is a unit of time
used for thousands of years
by the Jewish People
to calculate the lunar cycle.

Speed of light

The speed of light
in units of meters per chelek is:

1,000,000,000 [meters/chelek]

(Accuracy is 99.3%)

Wow!
That is a very round number!

This new proof is unique and irrefutable because it has all of these 7 points.
1. It shows an intelligent design that is not just intelligent in the sense of
the perfect functioning of nature. We are not claiming that life could not exist
without these ratios being close to round numbers, and in the range of 1, or 3, or 4.
2. It shows an intelligent numerical code.
3. The sheer number of coinciding ratios make the probability of this
code being by chance so close to zero, that it is indisputable.
4. This code has special meaning because it represents the name of G-d.
This shows that G-d literally signed His name into the physical architecture
of the solar system that He created.
5. The proof uses only simple math, and anyone with a high school education
can understand it, and validate that this proof is correct.
6. The code is not just conveyed written knowledge, but rather it is seen from
the physical size of the sun, moon, and earth. This means that the code was made
by the Creator of the sun, earth and moon. So it not only shows intelligence,
but it shows that only the powerful Creator of the world could have made this code.
7. There is more than one proof brought supporting this, there are a few
different ones in this video, and they all reinforce each other.

The old assumption that nothing is special has been debunked!

Since this is new revolutionary information, it needs to be looked at by the scientific and academic communities, and they need to reassess their entire belief system and assumptions.

Governments now need to reassess if it is ethical to use taxpayers money to teach in schools the curriculums that are still pushing the idea that science shows only randomness in the world, now that this has been proven false.

Donors need to reassess who they give their grants to.
Do people and institutions that still promote
this now debunked pseudo science
and cannot recognize an intelligent code
out of the seemingly randomness,
still deserve their support?

Public and private funding
needs to go to education, science, and research
that is not still stuck in a 200 year old bias
that forcefully tries to ignore the new scientific evidence showing
that the world has a Creator.

"You shall show proof of the truth to your friend."
This is a better translation than,
"You shall surely rebuke your neighbor." [Levictus 19, 17]
Because in Hebrew the word 'Hocheach'
literally means, "to prove".

Now it is possible to fulfil the commandment
to show your friends proof of the truth
by forwarding this video to them.

 SHARE

www.263672.com   www.Refuah.net

Zivi Ritchie, (BsEE) lives in Jerusalem, Israel. He is a Torah Observant Jew. He has a Bachelor of Science in Electronics Engineering from Machon Lev, Jerusalem College of Technology. He studied in Yeshiva Ohr Elchonan Chabad in Los Angeles, USA, and he studied Torah in Kollel in Israel for many years.

Zivi Ritchie is the CTO, Chief Technology Officer of Healables Digital Health, a high-tech startup. www.Healables.com

Zivi Ritchie is also the CEO of Refuah Institute, Online Professional CBT Coach and Counselor Training Programs in Accordance with Torah.

Zivi Ritchie's father, Professor Rabbi Joshua Ritchie, M.D., is the Founder and Dean of Refuah Institute, and he teaches many of the classes. www.Refuah.net

Zivi Ritchie

(Professor Rabbi Joshua Ritchie, M.D. researched and invented the treatment of preventing blindness, brain injury, and death in premature babies by administering to them vitamin E. Professor Rabbi Joshua Ritchie, M.D. did not attempt to make money from his discovery, but instead published his findings in the New England Journal of Medicine, so that everyone would be able to use this knowledge for free. Countless millions of people today owe their vision, their functioning minds, and their lives to him.)
Zivi Ritchie's mother is an artist and author of many spiritual books. www.lilianeritchie.com